Sana Aboulkacem
Rayène Ben Rhouma
Chekib Mazigh

# Monoclonal gammapathies: biological profile

Sana Aboulkacem
Rayène Ben Rhouma
Chekib Mazigh

# Monoclonal gammapathies: biological profile

## Monoclonal gammapathies: Biochemical exploration and epidemiological profile

ScienciaScripts

**Imprint**
Any brand names and product names mentioned in this book are subject to trademark, brand or patent protection and are trademarks or registered trademarks of their respective holders. The use of brand names, product names, common names, trade names, product descriptions etc. even without a particular marking in this work is in no way to be construed to mean that such names may be regarded as unrestricted in respect of trademark and brand protection legislation and could thus be used by anyone.

Cover image: www.ingimage.com

This book is a translation from the original published under ISBN 978-620-3-41180-5.

Publisher:
Sciencia Scripts
is a trademark of
International Book Market Service Ltd., member of OmniScriptum Publishing Group
17 Meldrum Street, Beau Bassin 71504, Mauritius
Printed at: see last page
ISBN: 978-620-3-36062-2

# Monoclonal gammapathies: Biochemical exploration and epidemiological profile

**EPS:** Serum protein electrophoresis

**IF:** Immunofixation

**EPGA:** Serum protein electrophoresis in agarose gel

**ECP:** Capillary serum protein electrophoresis

**Ig mc:** Monoclonal immunoglobulin

**Ig:** Immunoglobulin

**SDS:** Sodium Dodecyl Sulfate

**DNA:** Deoxyribonucleic Acid

**RNA:** Ribonucleic Acid

**Phi :** Isoelectric hydrogen potential

# List of figures

# List of tables

# Table of Contents

# Introduction

Clinical biochemistry is a field that focuses on the analysis of molecules present in body fluids such as blood, cerebrospinal fluid, urine. It allows us to follow the evolution of diseases and the effectiveness of treatments.

The analysis of serum proteins by electrophoresis and immunofixation which are two valuable examinations in many pathological situations such as monoclonal gammapathy, which is an abnormality defined by the presence of monoclonal immunoglobulin in the serum and/or urine. These immunoglobulins are proteins in human plasma that possess immune properties. They are synthesized in plasma cells of the lymphoid system cells formed in the spleen and lymph nodes. Monoclonal gammopathy is the proliferation of a clone of plasma cells that produces monoclonal immunoglobulin.

# Objective

The aim of this work is to study the epidemiological profile of this pathology and to explain the principles of the techniques of its chemical exploration.

The manuscript is composed of a first chapter devoted to a bibliographical study which presents some generalities on serum protein electrophoresis, immunofixation and monoclonal gammapathy.

The second chapter focuses on the experimental conditions in terms of materials and methods used.

The results and discussion are presented in the third and fourth chapters.

# CHAPTER 1
## Bibliographic synthesis

### 1. Electrophoresis

### 1.1 Historical background

As early as the end of the 18th century, Michael Faraday presented his electrolysis work which is based on the laws of electrostatics and electricity. (1)

A century later, in 1892, Linder S.E. and Picton exploited the ancient research of electrolysis for the separation of charged particles: it is the phenomenon of electrophoresis which allows the displacement of macro ions, particles and bacteria under the action of an electric field, in aqueous solution or in other solvents. (2)

In 1937, the Swede Arne Wilhelm Kaurin Tiselius applied this technique in the separation of proteins from serum using the so-called free solution or mobile boundary method, which would earn him the Nobel Prize in 1948. Since then, many electrophoresis separation processes have appeared. (2)

In 1939, P. König and D Von Klobusitzky succeeded in separating the constituents of snake venom using the new technique: electrophoresis on paper.

In 1959, Raymond and Weintraub reported on polyacrylamide gel electrophoresis.

In 1969, Beber and Osborn introduced the denaturing agent SDS (Sodium Dodecyl Sulfate) to evaluate the purity and determine the molecular weight of proteins. (1)

A few years later, the introduction of capillary electrophoresis caused a real revolution in the separation of biochemical molecules. It became one of the most developed analytical techniques from the 80's.(2)

Today this technique has become an indispensable tool in many laboratories, both for research and industry and for medical sciences. At the same time, the appearance of electrophoresis automatons has allowed a better standardization of the different stages of the separation, in particular the method of sample deposition and the control of migration parameters such as voltage, temperature and migration time. (1)

## 1.2.	General information on electrophoresis

Electrophoresis is certainly the technique most commonly used by biochemists and biologists to fractionate macromolecules (proteins, nucleic acids)(3).

Principle :
An ion is any charged particle that can move under the influence of an electric field. These particles migrate towards their respective electrodes: anions (negatively charged) migrate towards the anode (positive pole) while cations (positively charged) migrate towards the cathode (negative pole). (4)

Note :
The migration is done by a well determined speed called migration speed which determines the nature of the components to be separated. This depends not only on the charge of the proteins, their size and shape but also on experimental conditions such as pH, composition of the solutions and electrophoretic support.

The revelation is made after staining the proteins with a dye such as amidoschwartz. (5)

## 1.3.	The different types of electrophoresis

Separation techniques based on this principle are called electrophoretic separation methods and can be divided into three main types: zone electrophoresis, liquid vein electrophoresis and capillary electrophoresis. (2)

### 1.3.1.	Zone electrophoresis

It is the set of electrophoretic techniques whose separation principle is based on the difference of mobility.
In this case the concentration of the electrolytes, which is high compared to that of the ionic species in the sample, allows a relatively constant pH and a uniform potential gradient to be maintained. Under the action of an electric current, the ions in the sample migrate at a constant rate, resulting in a flow of ions from the support electrolyte followed by a flow of ions from the sample. (2)

1.3.1.1 Electrophoresis   on a paper        support

The general principle of this technique is based on the use of a suitably dimensioned and stretched paper web. The end of this strip is extended in a tank containing the electrolyte solution which must ensure the constancy of the ionic strength and, that of the pH. On each side (left and right) of the paper, two electrodes are placed to establish a potential difference. The solution to be separated is deposited on a point on the upper side. (2)

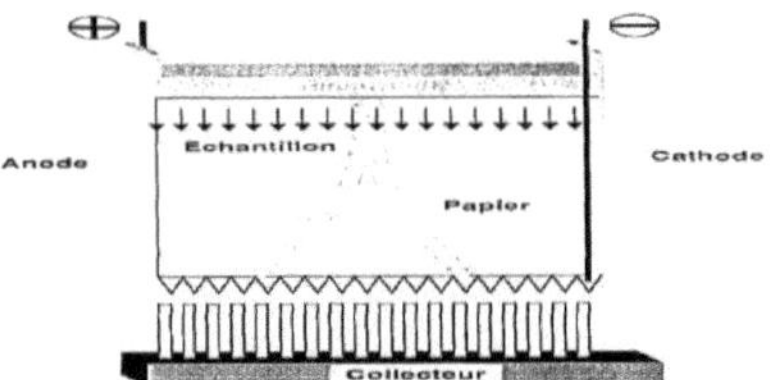


*FIGURE 1. **REPRESENTATION OF THE SEPARATION BY ELECTROPHORESIS ON A PAPER SUPPORT. (2)***

1.3.1.2.  Cellulose acetate membrane electrophoresis

It is a technique discovered by Kohn in 1957. It is simpler than electrophoresis on paper. Thanks to its transparency and uniformity of porosity, the cellulose acetate membrane has replaced paper. The cell is composed of two compartments which are filled with a buffer solution where the electrodes are placed: cathode and anode. These two compartments are separated by a membrane. (2)

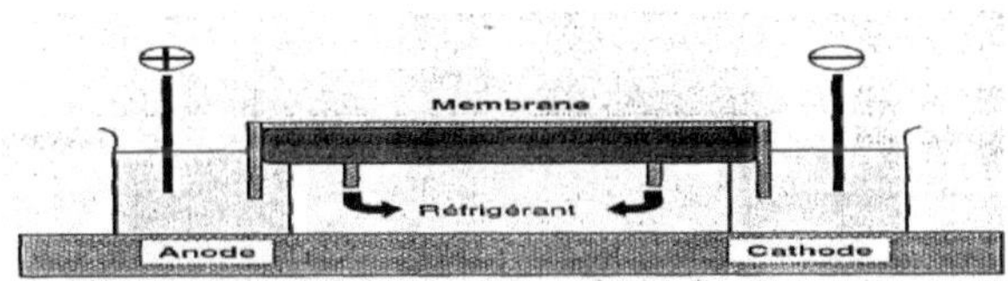


*FIGURE 2. **REPRESENTATION OF SEPARATION BY ELECTROPHORESIS ON CELLULOSE ACETATE MEMBRANE.***

1.3.1.3. Gel electrophoresis

This technique was proposed by Smithies in 1955. It uses a partially hydrolyzed starch that allows the migration of large molecules. The proteins to be separated are inserted into a slit in the gel. After migration, the gel is cut through its thickness and the proteins are revealed by one of the usual dyes.

In 1960 S. Raymond and WANG (1960) used a new acrylamide-based gel. In this gel there is a mesh whose size can be controlled, so the speed of migration depends not only on the charge density but also on the size of the molecule.

The instrumentation consists of a tank where the gel is placed, sandwiched between two glass or acrylic plates and the two electrodes connected to an electric current generator.

Note :

The choice of gel depends on the nature of the molecules to be separated:

**TABLE 1. *NATURE OF THE SEPARATION GEL ACCORDING TO THE MOLECULE TO BE SEPARATED***

| Type of gel | starch gel | polyacrylamide gel | agarose gel |
| --- | --- | --- | --- |
| **Molecules to be separated** | Iso-enzymes | proteins or nucleic acids of short sequence. | proteins, DNA, RNA |

1.3.2. Liquid vein or free electrophoresis

It is a technique discovered by Tiselius in 1937. The ionic species to be separated in solution, fill the lower part of a U-tube. The upper part is filled with the buffer solution. If an electric current is applied to the system, the anionic species migrate in the direction of the anode and, after some time, a partial separation occurs. The results are shown by optical methods (UV absorption, refractive index, fluorescence...). This method is used to measure electrophoretic mobility and to check the purity of proteins. (2)

1.3.3 Capillary electrophoresis

Capillary electrophoresis was invented by Hjerten, Everaerts and Keulemans in the late 1960s.

Capillary electrophoresis is performed in a fused silica tube covered with a polyamide layer, 20 to 200 cm long and with an internal diameter of 20 to 200 μm. The capillary, enclosed in a thermostatic system, is filled with buffer and immersed in two tanks containing the same solution. Each tank is connected to an electrode connected to a current generator. The apparatus also includes a detection system, a UV-visible spectrophotometer or a fluorometer. (6)

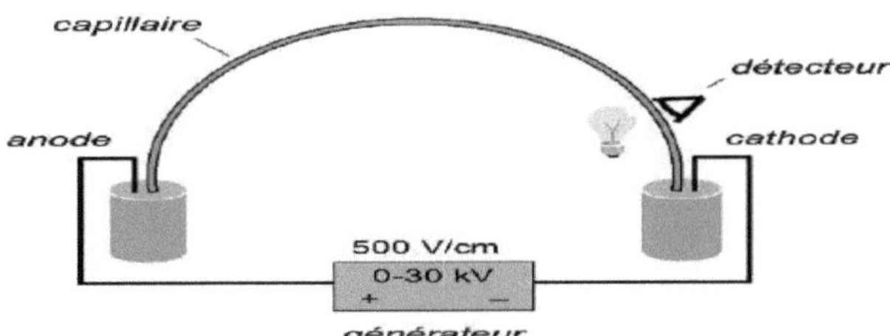


*FIGURE 3. SCHEMATIC REPRESENTATION OF A CAPILLARY ELECTROPHORESIS APPARATUS. (6)*

## 1.4  Serum protein electrophoresis

Serum protein electrophoresis (SPE) is a technique routinely performed by medical analysis laboratories. It allows the separation of proteins contained in a serum subjected to the action of an electric field and results in the individualization of 5 or 6 fractions by indicating their percentages. (7)

There are two methods that are suitable for the analytical fractionation of serum proteins: agarose gel PSA and liquid vein PSA in fused silica capillary. (8)

1.4.1. Steps of serum protein electrophoresis

- Migration :

The electrophoretic separation depends on the isoelectric point (pHi) of the molecule in relation to the pH of the migration buffer. At basic pH, the negatively charged protein will migrate to the anode and at acidic pH, the positively charged protein will migrate to the cathode. (9)


- The revelation :

It is done :

- or by dyes: in the case of EPS on agarose gel, acid violet, Coomassie blue or amidoschwartz is used.

-either by antisera, followed by staining: by immunoelectrophoresis or immunofixation (antibodies specific to the molecule). (9)

- Electrophoretic integration :

This step requires either :

- A densitometry: in this case the coloration is proportional to the concentration of the molecules to be separated.

- An optical density: UV reading at 200 nm which is the absorption wavelength of the peptide bonds(9)

### 1.4.2. Blood proteins and different fractions

Serum proteins play several enzymatic, immune and transport functions. They are involved in the acid-base balance and the maintenance of oncotic pressure, as well as playing an important nutritional role. They are mainly synthesized by hepatocytes but also by lymphocytes and plasma cells.(7)Whatever the method used, the separation of proteins contained in a serum subjected to the action of an electric field results in the individualization of 5 or 6 fractions. Going from the anode to the cathode we find the albumin fraction, then the globulin fractions $\alpha1$, $\alpha2$, ß (or ß1 and ß2) and finally $\gamma$ (gamma). (8)

***TABLE 2. THE DIFFERENT FRACTIONS OF SERIAL PROTEINS AND THEIR ELECTROPHORETIC MIGRATION (8)***

| Fraction | Composition and electrophoretic profile |
|---|---|
| **Albumine** | Albumine |
| **Alpha 1** | the orosomucoid<br>α1antitrypsine |
| **Alpha 2** | α2macroglobuline<br>haptoglobin |
| **Beta** | Hemopexin<br>Fraction C3 of the<br>transferrin complement<br>(TRF)<br>IgA fraction |
| **Gamma** | IgG, IgA IgM |

1.4.3. The applications of EPS :

Serum protein electrophoresis (SPE) is an orientation test with very broad indications.
 The EPS is used to highlight :

An inflammatory profile, nephrotic syndrome, hypo- or hyper-gamma globulinemia,
undernutrition, or monoclonal immunoglobulin(10)

1.4.4. The different electrophoretic profiles of the EPS :

The reading of the results by densitometry gives different profiles :

1.4.4.1. Normal profile

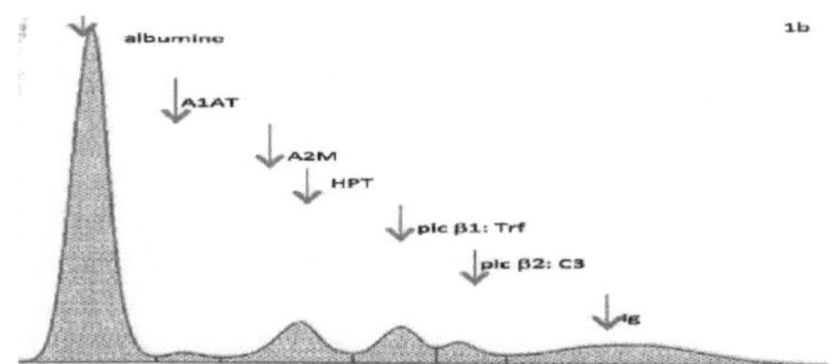


**Figure 4.** Normal profile of serum protein electrophoresis

*TABLE 3.THE NORMAL COMPOSITION OF SERUM PROTEINS*

| Name | % | g / l |
|---|---|---|
| **Albumine** | 55 - 65 % | 36 - 50 g /l |
| **α1 - globulins** | 1 - 4 % | 1 - 5 g /l |
| **α2 - globulins** | 6 - 10 % | 4 - 8 g/l |
| **β- globulins** | 8 - 14 % | 5 - 12 g /l |
| **γ - globulins** | 12 - 20 % | 8 - 16 g /l |

Serum proteins are separated into 5 fractions :

- A fraction near the anode consisting of albumin.
- The following 3 fractions α1, α2, ß.
- A large fraction located near the cathode and consisting of immunoglobulins: this is
  the fraction γ. (8)

1.4.4.2. Pathological profiles of EPS

A pathological profile corresponds to the variation of one or more protein fractions due to an abnormality.

*TABLE 4. EPS PATHOLOGICAL PROFILES*

| Type of anomaly | Quantitative anomaly<br><br>nephrotic syndrome | Qualitative anomaly<br><br>monoclonal gammopathy |
| --- | --- | --- |
| **Description** | Nephrotic syndrome is associated with high proteinuria, hypoalbuminemia and hypoprotidemia. It is due to an alteration of the renal glomeruli, (diabetes, connectivitis and certain infections or toxicities). This disease can be complicated by high blood pressure or a renal insufficiency. (11) | Monoclonal gammopathy is a condition that results in a monoclonal immunoglobulin (Ig mc) spike(12) . |
| **Electrophoretic profile (EPS)** | Diagnosis is based on the clinic by analysis of urinary and serum proteins. Following a massive urinary leakage of proteins which leads to a decrease in plasma proteins we notice : <br>- hypoalbuminemia of less than 30 g/L. <br>- lower hypoprotidemia to 60 g/L. (11) | The serum protein electrophoresis technique allows the evaluation of this anomaly, the Ig mc migrate as a narrow and homogeneous band located in the gamma region and more rarely in the beta and alpha 2 regions (12). |

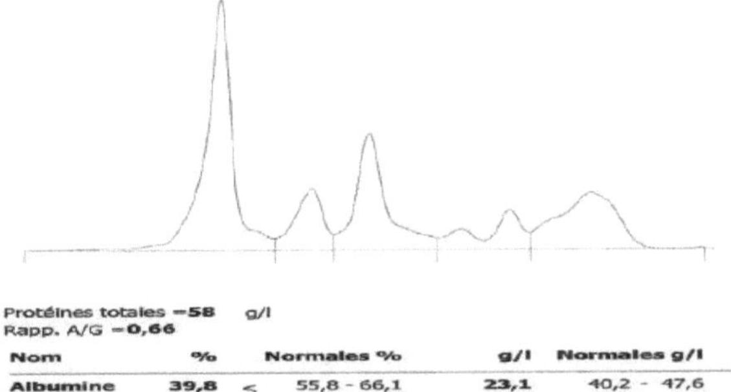

Protéines totales =58    g/l
Rapp. A/G =0,66

| Nom | % | | Normales % | g/l | Normales g/l |
|---|---|---|---|---|---|
| Albumine | 39,8 | < | 55,8 - 66,1 | 23,1 | 40,2 - 47,6 |
| Alpha 1 | 10,2 | > | 2,9 - 4,9 | 5,9 | 2,1 - 3,5 |
| Alpha 2 | 21,5 | > | 7,1 - 11,8 | 12,5 | 5,1 - 8,5 |
| Beta | 9,1 | | 8,4 - 13,1 | 5,3 | 6,0 - 9,4 |
| Gamma | 19,4 | > | 11,1 - 18,8 | 11,3 | 8,0 - 13,5 |

*FIGURE 4. **SERUM PROTEIN ELECTROPHORESIS PROFILE CHARACTERISTIC OF A NEPHROTIC SYNDROME. (11)***

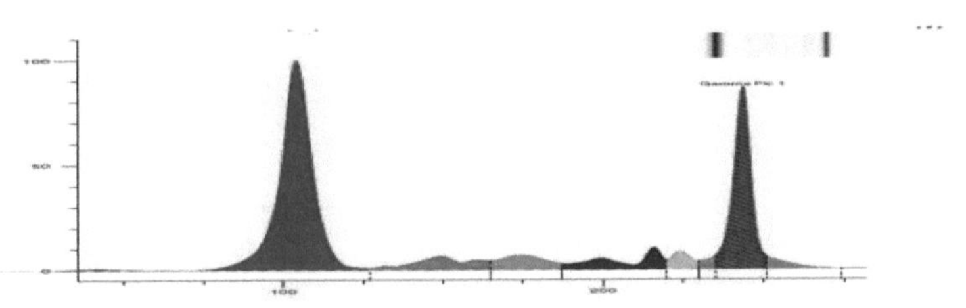


*FIGURE 5. **ELECTROPHORETIC PROFILE OF MONOCLONAL GAMMAPATHY . (13)***

## 1.    Immunofixation (IF)

### 2.1.Definition

Immunofixation (IF) is a technique used for typing monoclonal immunoglobulins detected as peaks in PSA(14)

### 2.2.  Principle

IF is a gel immunoprecipitation technique. This gel contains electrophoretic migration tracks on which the samples are deposited. After electrophoretic migration of the serum constituents, the different tracks are incubated in the presence of specific antisera. If Ig is present, immunoprecipitation occurs. After washing, protein dyes are used which allow us to visualize the reaction. Finally the results are commented in the presence of a biologist.

### 2.3.  Application

The combination of an EPS, an IF with the free light chain assay allows us to detect a monoclonal component of any etiology: multiple myeloma, Waldenstrom's disease, plasmocytoma, light chain disease. (15)

### 2.4.  Interpretation

The immunofixation profile can tell us about different abnormalities:
- The presence of a wide, diffuse band points to the presence of polyclonal immunoglobulins.
- The presence of a narrow band at the same level of migration with an anti heavy chain immuno-serum (gamma, alpha or mu) and an anti light chain immuno-serum (anti Kappa or anti Lambda) indicates the presence of a monoclonal immunoglobulin. This band is also found on the first control track.
- The presence of a narrow band with a light chain immuno-serum with no bands with the heavy chain antisera should incite to perform a second IF using an anti-delta and anti-epsilon antisera to look for Ig D or Ig E (rare entities).
- The presence of a narrow band with a non-striped heavy chain immune serum with light chain immune serums evokes a heavy chain disease(15).

# 3. Monoclonal gammopathy

## 3.1. Definition

Monoclonal Gammapathy was first described in 1978 (16). It is a condition defined by the presence of a peak in the region of gamma globulins or sometimes in beta globulins and very rarely in alpha globulins explained by the presence in serum or urine of an abnormal amount of monoclonal immunoglobulin (12). This peak is sharp and narrow-based. (13)

## 3.2. Nature of immunoglobulin

The nature of the immunoglobulin: IgG, IgA, IgM, IgD and its monotype: same heavy chain and same light chain must be confirmed by several techniques: IF or immunoelectrophoresis.

*TABLE 5. DISTRIBUTION OF HEAVY AND LIGHT CHAINS . (17)*

| heavy chain | light chain |
| --- | --- |
| **IgG, 55-65% IgG, 55-65% IgG, 55-65% IgG** | Independently of the nature of the heavy chain we find : |
| **IgA, 10-15%.** | -2 /3 of cases kappa chain |
| **-IgM, 15 to 20%.** | -1/3 of the cases lambda chain |

## 3.3. Causes of monoclonal gammopathies

The pathologies that can lead to monoclonal gammapathy are :

3.3.1 Hematologic malignancies

-Multiple myeloma: this is a malignant disease of the bone marrow characterized by an abnormal proliferation of a plasma clone.
-Waldenstrom's disease.
-Lymphoma B: it is a lymphoproliferative disorder of the B cells producing this Immunoglobulin.

3.3.2. Non-malignant diseases

In this case we are talking about chronic (biliary, urinary) or acute infections (infectious mononucleosis, cytomegalovirus or AIDS virus infections), autoimmune diseases (rheumatoid arthritis, systemic lupus, cirrhogenic liver diseases, glomerular nephropathies).

Certain constitutional immune deficiencies and hematopoietic stem cell transplantation. (13)

The diagnosis of monoclonal gammopathy of undetermined significance is retained if the other causes are eliminated.

## 3.4. Complication of monoclonal gammapathies

Monoclonal gammopathies can lead to complications such as :

Amyloidosis: which is a group of diseases diagnosed histologically by the presence of protein deposits in tissues (kidneys, heart, nerves, liver...) that can cause a failure of these organs such as glomerular damage, kidney failure, heart failure, ....

Plasma hyperviscosity syndrome: most often associated with elevated IgM levels during Waldenström's macroglobulinemia. It associates neurosensory and hemorrhagic signs, with headaches, dizziness, hearing loss, visual disturbances, torpor and drowsiness, and sometimes convulsions.

Cryoglobulinemia: which refers to the presence of immunoglobulins, or a combination of immunoglobulins and complement proteins, in the blood that precipitate when cold. Most often asymptomatic, they can lead to cutaneous manifestations (purpura, Raynaud's phenomenon, even necrosis of the extremities), polyarthralgia, glomerular nephropathies. (18)

## 3.5. Epidemiology

Several scientific researches on monoclonal gammopathy have studied the epidemiological, clinical and biological profile: age, sex, types of heavy and light chains, monoclonal component levels, blood glucose, blood count, creatinine, myelogram....
This disease is more common in men than in women, its median age is around 72 years. Monoclonal gammopathy is twice as common in the black population, especially those of Caribbean origin, while it is rare in the Asian continent. Prevalence evolves with age, increasing to 1% in the population at age 50, 3% at age 70 and 5-10% after age 80(13).

# CHAPTER 2
## Materials and Methods

## 1. Hardware :

This is a retrospective study. We have collected all the samples arriving at the EPS bench of the clinical biochemistry laboratory in the main military training hospital in Tunis during the year 2019.

### 1.1. Equipment :

* the HYDRSIS 2 scan (Sebia) * Micropipette

* Centrifuge* Fine                                    filter papers

* Densitometer* Applicators

*Unicel DXC*C800 Beckman Coulter* Antiserum  strips

* Thick filter papers

*FIGURE 6. THE HYDRASIS 2 SCAN MACHINE*

### 1.2. Products :

The analysis of blood samples requires the use of the products presented in Table 6.

*TABLE 6.PRODUCTS USED FOR EPS AND IF*

| For the EPS technique | |
|---|---|
| **Product** | Constituent |
| **Agarose gel** | Agarose 8g /l /Tri-barbital buffer (pH=9.2) |
| **Stamped wicks** | Tri-barbital buffer (pH=9.2)/Sodium azide |
| **Coloring diluent (1 bottle of 60 ml)** | Concentrated acid solution |
| **Amidoschwarz dye (1 bottle of 20 ml)** | Concentrated blue solution: Amidoschwarz |
| **Decolorant (1 bottle of 100 ml)** | Citric acid 0.5g/l |

| For the immunofixation technique | |
|---|---|
| **Product** | Constituent |
| **Washing solution** | a buffer of pH=8.7 |
| **Acid violet dye (1 vial of 75 ml)** | Concentrated solution: acid violet |
| **Fixer (1 bottle of 2 ml)** | Acid solution of pH=2 with a specific color |
| **Ig tolales anti-gamma chains (1 vial of 2 ml)** | Total mammalian immunoglobulins anti gamma chain |
| **Anti-alpha-chain total Ig (1 vial of 2 ml)** | Total mammalian immunoglobulins anti-alpha chain |
| **Anti-mu-chain total Ig (1 vial of 2 ml)** | Mammalian total immunoglobulins anti mu-chain |
| **Total Ig anti kappa chain (1 bottle of 2 ml)** | Total mammalian immunoglobulins anti-kappa chain |
| **Total Ig anti-chain lambda (1 vial of 2 ml)** | Total mammalian immunoglobulins against lambda chain |

**Preparation of the dye amidoschwarz :**

The concentrated amidoschwarz dye is a viscous solution. In order to obtain a perfect reconstitution of the dye, the following protocol must be followed:

- Add 15 ml of coloring diluent to the concentrated amidoschwarz bottle.
- Close the bottle carefully.
- Shake the bottle very vigorously for at least 5 seconds.
- Pour the resulting solution into the staining solution preparation vessel.
- Repeat this operation two or three times.
- Pour the rest of the diluent into the preparation container of the staining solution.
- Make up to 300 ml with distilled or demineralized water.
- Shake this solution perfectly.
- The dye is ready to use

**2.     Methods :**

**2.1. Pre-analytical phase :**

2.1.1.  Prescription and registration :

This phase begins with the prescription of the analysis. The prescription must indicate: the patient's name, the department where he or she is hospitalized, and clinical information. It is followed by the registration of the request to the laboratory.

2.1.2 Sampling

The sample is taken from the vein in the fold of the elbow in a fasting patient, in a supine position with the forearm in supination. The blood is collected in a tube without anticoagulant (dry tube).

The samples are left to rest for one hour to coagulate. After centrifugation at 3000 rpm for 5 minutes, sera are obtained for electrophoresis.

2.1.3 Transport and routing :

The routing of the sample must respect the appropriate conditions (temperature, speed of transport to the laboratory, etc.).

Analyses are performed on fresh samples that are not frozen (stored for a maximum of 8 days at a temperature of 2°C to 8°C) or frozen within 24 hours after sampling if the analysis of the specimen is to be delayed.

## 2.2. Analytical phase :

PSA and immunofixation are performed according to the experimental protocol described below:

### 2.2.1. Total serum protein assay :

The total protein determination is based on a biuret colorimetric method adapted to the Beckman Coulter Unicel DXC 800. The Gornall reagent (composed of copper sulfate) is used in this case. The intensity of the staining is proportional to the number of peptide bonds. The maximum absorption of these bonds is between 530-550 nm.

### 2.2.2. Procedure for serum protein electrophoresis

Sera are used as samples to avoid the appearance of a supernumerary fibrogenic peak. The use of hemolyzed samples is also avoided.

-Sample Preparation :

- o Place 10 µl of the samples in the applicator wells.
- o Place the applicators in the wet chamber for 5 minutes.

- Preparation of the migration :

This step begins with the fixation of the buffered wicks which act as a buffer reservoir and ensure contact between the gel and the electrodes.

- o Quickly remove excess liquid from the surface by brushing the gel through the filter paper.
- o Place 120 µl of distilled or demineralized water on the migration tray in the lower third of the screen printing frame.
- o Place the gel on the tray against the clip, inside the screen-printed frame.
- o Give a concave shape to the gel and roll it out on the tray until it contacts the water drop which must be distributed along the entire length of the gel by eliminating trapped air bubbles.
- o For the analysis of 30 samples, the applicators are placed in positions number 3 and 9 on the applicator holder.
- o Close the cover of the migration module and start the sequence immediately.

- Preparation of gel treatment sequences: staining

- o Place the gel film on the film holder, gel side towards the operator.
- o Introduce it into the treatment / gel staining module
- o Select the coloring program.

- End of gel treatment :

In this step, the dry gel is removed and if necessary, the posterior part (plastic support) is cleaned with a damp padded paper for a good densitometer/scanner reading at 570 nm.

- Interpretation :

If the electrophoretic profile evokes monoclonal gammapathy (presence of a peak in the gamma globulin zone and rarely in the beta and alpha 2 globulin zones), immunofixation is carried out.

### 2.2.3. Procedure for immunofixation

This technique is used to identify the immunoglobulins present in the sample: nature of heavy chains and light chains. It is carried out using monospecific antisera for the identification of monoclonal bands detected by EPS. The protocol consists mainly of 4 steps:

- o Electrophoretic separation of proteins in agarose gel: this step is performed during serum protein electrophoresis.
- o Fixation and immunoprecipitation: application of fixative and antisera to the gel at the migration tracks. The fixative and antisera diffuse into the gel.
  The fixative precipitates all proteins and the antibodies precipitate the corresponding antigens.
- o Removal of non-precipitated proteins by pumping and washing (by the washing solution of pH=8.7). Precipitated proteins remain trapped in the gel.
- o Protein staining and comparison of the position of immunoprecipitated bands with that of abnormal bands observed in serum protein electrophoresis.


Note :

Some proteins and light chains tend to polymerize and form aggregates. In this case, the sample (serum) is treated with a solution of Beta-2 mercaptoethanol. For this purpose these steps must be followed:

- Dilution 1/10th (9 volume of serum + 1 volume Beta-2 mercaptoethanol pre-diluted 1/10th with 0.9% NaCl).

- Allow to incubate for 10 minutes and then dilute the serum with the immunofixation-specific diluent.

## 2.3. Post-analytical phase

The post-analytical phase corresponds to the interpretation of the immunofixation results.

# CHAPTER 3
## Results

### 1. Description of identified patients

We have collated specimens arriving at the EPS service with a profile suggestive of monoclonal gammapathy.

We included 56 patients for whom serum and urinary immunofixation (IF) was performed. Of these 56 IFs, 36 (64%) were positive:

*TABLE 7.BREAKDOWN OF IMMUNOFIXATION RESULTS (IF)*

| | |
|---|---|
| **Total number of patients who had IF** | 56 |
| **Normal FI** | 20 |
| **Pathological IF** | 36 |
| **Percentage of affected patients tested by IF** | 64,29% |

The 36 patients were divided according to several criteria: gender, age, and type of myeloma.

### 2. Distribution by gender

The study included 36 patients: 24 men and 12 women. The sex ratio H /F was 2 with a male predominance.

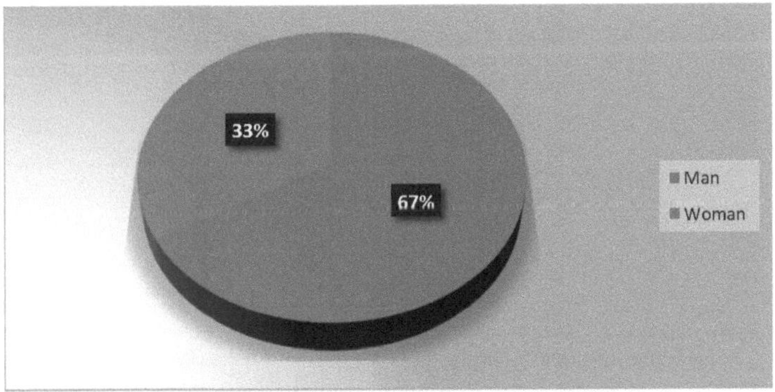


***Figure 7***. The distribution of monoclonal gammapathies by gender

### 3. Age distribution

The mean age of patients with a positive FI was 60 years with extremes ranging from 38 to 86 years.

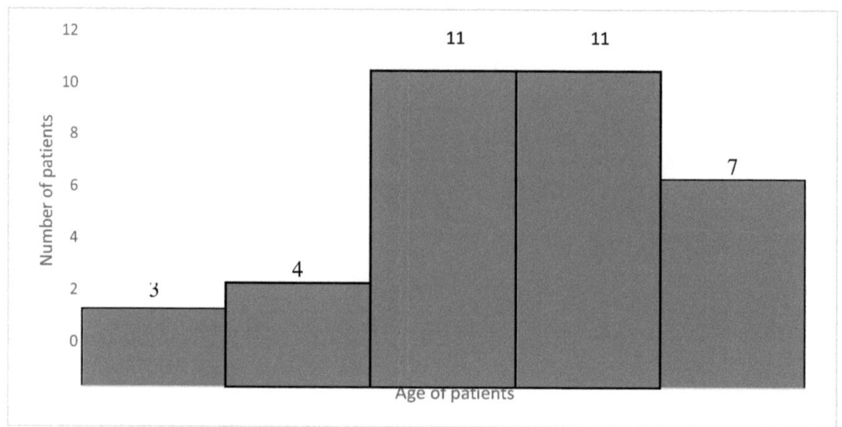


*FIGURE 8. AGE DISTRIBUTION OF MONOCLONAL GAMMOPATHIES CASES*

### 4. Electrophoretic profiles

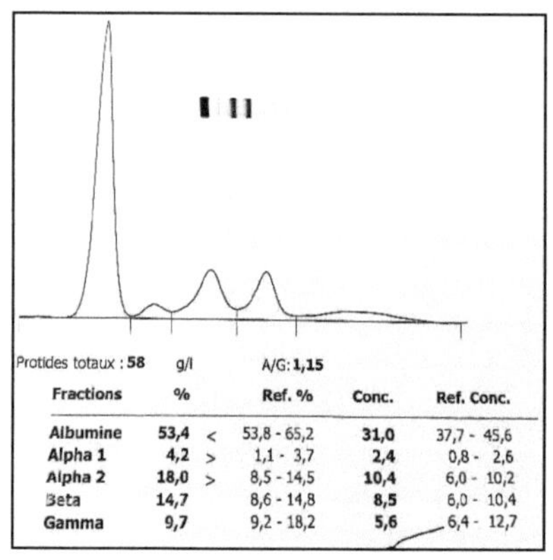

Protides totaux : **58**  g/l          A/G: **1,15**

| Fractions | % | | Ref. % | Conc. | Ref. Conc. |
|---|---|---|---|---|---|
| Albumine | 53,4 | < | 53,8 - 65,2 | 31,0 | 37,7 - 45,6 |
| Alpha 1 | 4,2 | > | 1,1 - 3,7 | 2,4 | 0,8 - 2,6 |
| Alpha 2 | 18,0 | > | 8,5 - 14,5 | 10,4 | 6,0 - 10,2 |
| Beta | 14,7 | | 8,6 - 14,8 | 8,5 | 6,0 - 10,4 |
| Gamma | 9,7 | | 9,2 - 18,2 | 5,6 | 6,4 - 12,7 |

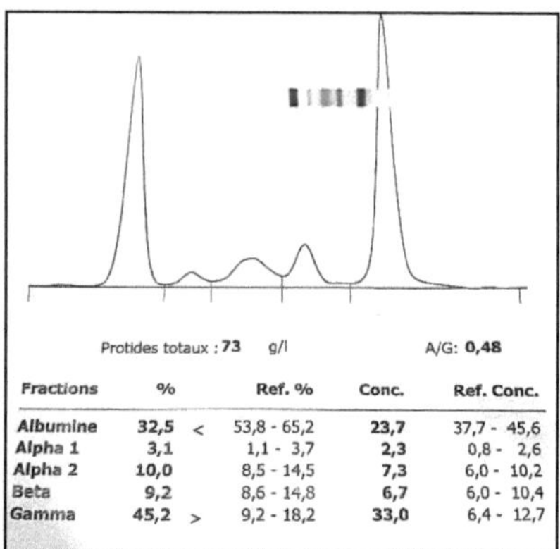

Protides totaux : **73**  g/l          A/G: **0,48**

| Fractions | % | | Ref. % | Conc. | Ref. Conc. |
|---|---|---|---|---|---|
| Albumine | 32,5 | < | 53,8 - 65,2 | 23,7 | 37,7 - 45,6 |
| Alpha 1 | 3,1 | | 1,1 - 3,7 | 2,3 | 0,8 - 2,6 |
| Alpha 2 | 10,0 | | 8,5 - 14,5 | 7,3 | 6,0 - 10,2 |
| Beta | 9,2 | | 8,6 - 14,8 | 6,7 | 6,0 - 10,4 |
| Gamma | 45,2 | > | 9,2 - 18,2 | 33,0 | 6,4 - 12,7 |

A flattening in the gamma globulin area

**Hypogammaglobulinemia**

Presence of a peak in the gamma globulin area.

**Hypergammaglobulinemia**

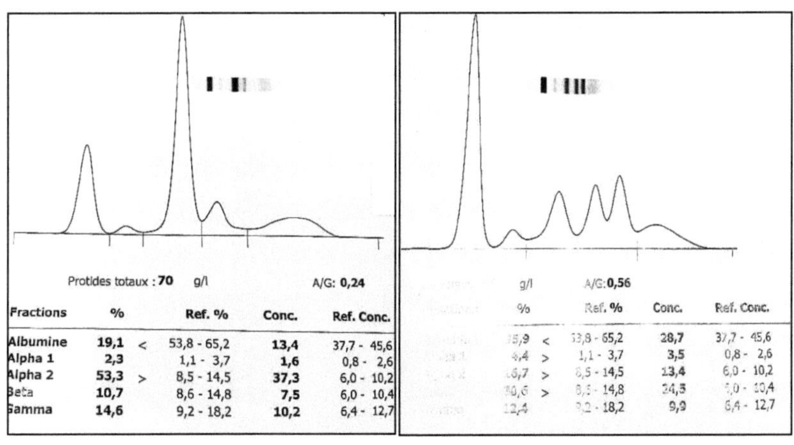



| Fractions | % | | Ref. % | Conc. | Ref. Conc. |
|---|---|---|---|---|---|
| Albumine | 19,1 | < | 53,8 - 65,2 | 13,4 | 37,7 - 45,6 |
| Alpha 1 | 2,3 | | 1,1 - 3,7 | 1,6 | 0,8 - 2,6 |
| Alpha 2 | 53,3 | > | 8,5 - 14,5 | 37,3 | 6,0 - 10,2 |
| Beta | 10,7 | | 8,6 - 14,8 | 7,5 | 6,0 - 10,4 |
| Gamma | 14,6 | | 9,2 - 18,2 | 10,2 | 6,4 - 12,7 |



| | % | | Ref. % | Conc. | Ref. Conc. |
|---|---|---|---|---|---|
| | 15,9 | < | 53,8 - 65,2 | 28,7 | 37,7 - 45,6 |
| | 4,4 | > | 1,1 - 3,7 | 3,5 | 0,8 - 2,6 |
| | 16,7 | > | 8,5 - 14,5 | 13,4 | 6,0 - 10,2 |
| | 50,6 | > | 8,5 - 14,8 | 24,5 | 6,0 - 10,4 |
| | 12,4 | | 9,2 - 18,2 | 9,9 | 6,4 - 12,7 |

| | |
|---|---|
| Presence of a peak at the level of alpha 2 globulins.<br><br>**Hyperalpha-2-globulinemia** | Presence of a peak at the level of beta globulins.<br><br>**Hyperbetaglobulinemia** |

***FIGURE 9. ELECTROPHORETIC PROFILES OF MONOCLONAL GAMMAPATHIES***

### 5. Breakdown by isotype

Serum IF typing showed a predominance of IgG: 18 (50%) were IgG (8 Lambda and 10 Kappa IgG), 10 (28%) were IgA (5 Lambda and 5 Kappa IgA), 1 (2.78%) were Kappa IgM and 3 (8.33%) were light chain (2 Kappa and 1 Lambda).

Four oligoclonal profiles were recorded at serum IF of which 2 were IgG Kappa and IgG Lambda, 1 was IgG Lambda and IgA Kappa and 1 was IgG Kappa and IgG Lambda.

The kappa to lambda ratio was 1.25. The kappa to lambda ratio was 1.25.

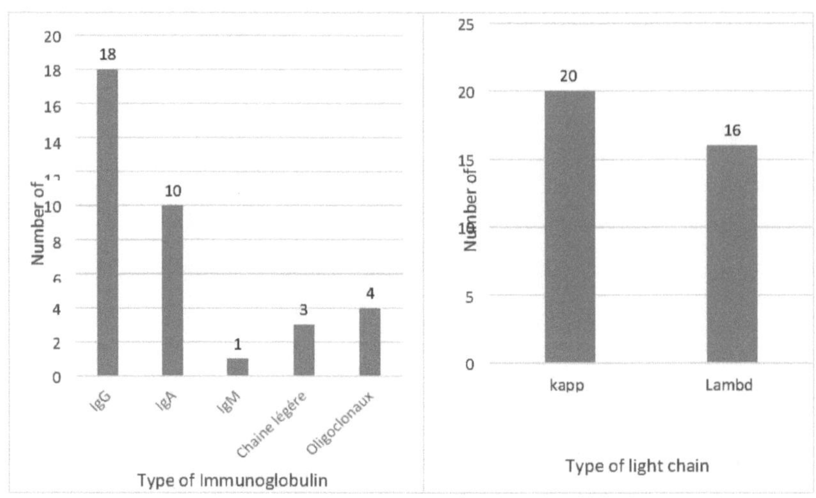



*FIGURE 10.* ***DISTRIBUTION OF ISOTYPES***

Bence Jones proteinuria was positive in 19 cases (52.78% of the cases): 10 lambda and 9 Kappa.

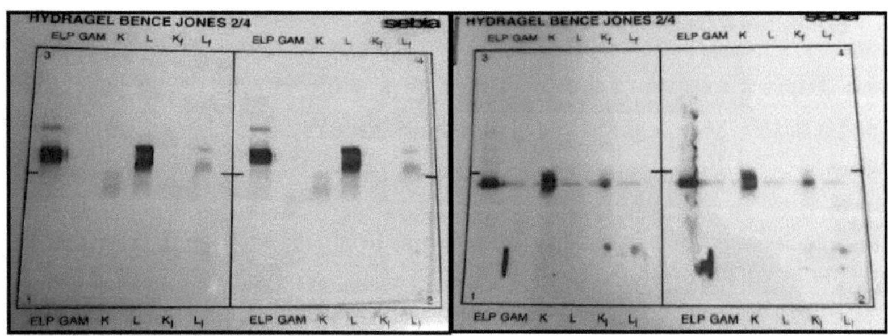



| Proteinuria of Bence Jones lambda | Proteinuria of Bence Jones kappa |

*FIGURE 11.* ***PROTEI NURIE DE BENCE JONES***

| Type of myeloma | Results |
| --- | --- |
| • IgG monoclonal immunoglobulin: 18 cases | o Lambda IgG immunoglobulin: 8 cases 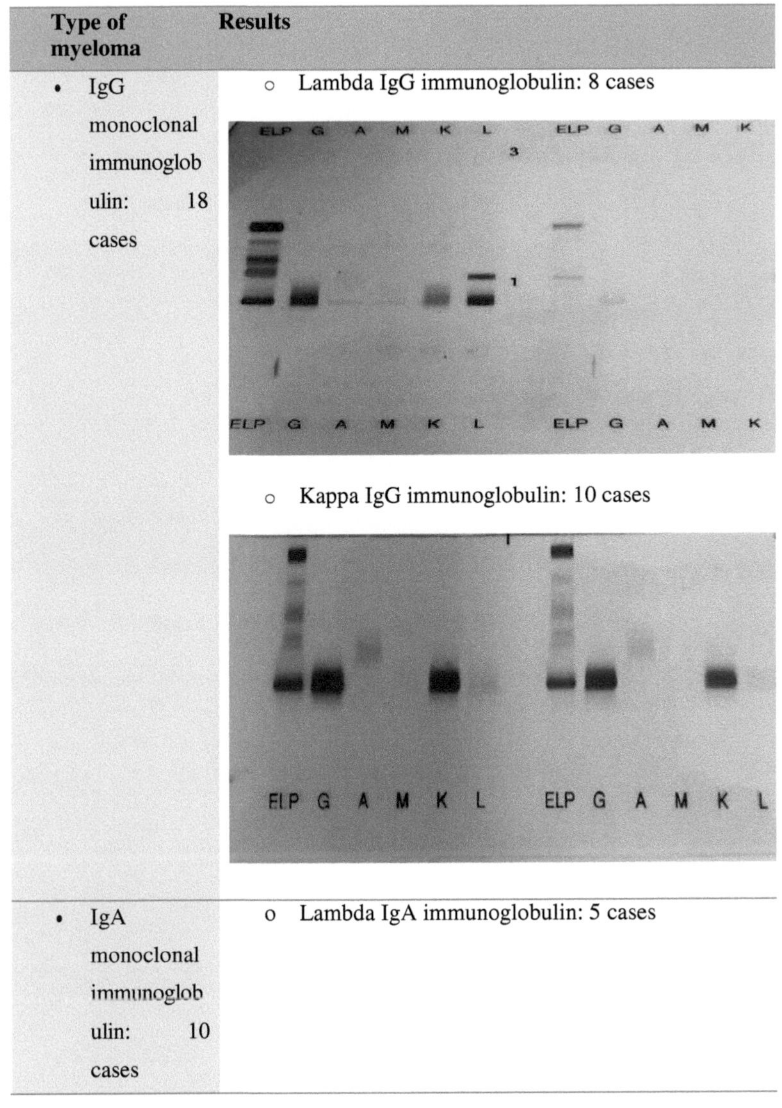<br>o Kappa IgG immunoglobulin: 10 cases |
| • IgA monoclonal immunoglobulin: 10 cases | o Lambda IgA immunoglobulin: 5 cases |

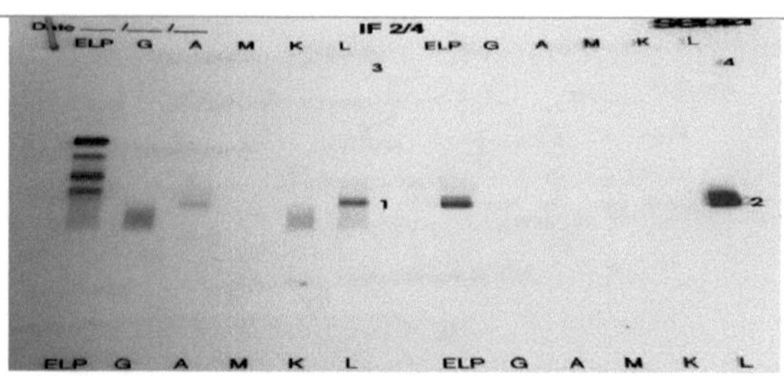

o   Kappa IgA immunoglobulin: 5 cases

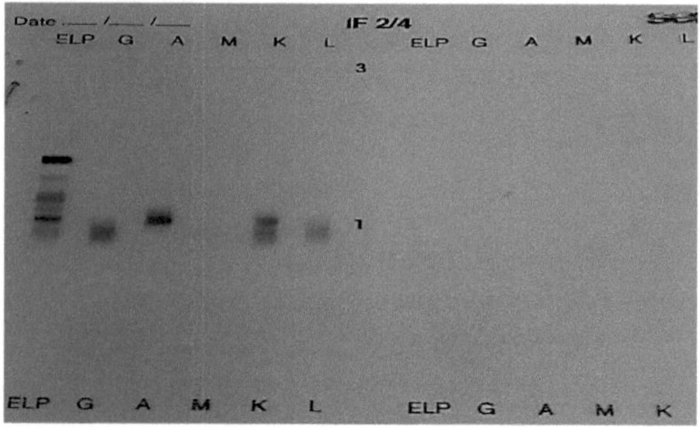

- IgM monoclonal immunoglob ulin: 1 case only

o   Kappa IgM Immunoglobulin: 1 case

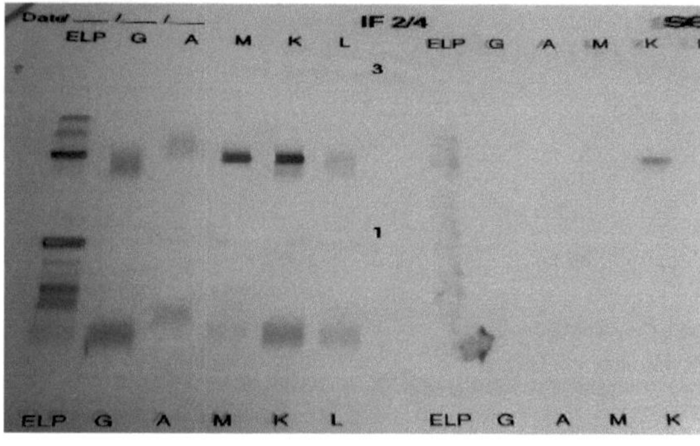

| • Light chain monoclonal immunoglobulin : 3 cases (2 Kappa and 1 Lambda) | 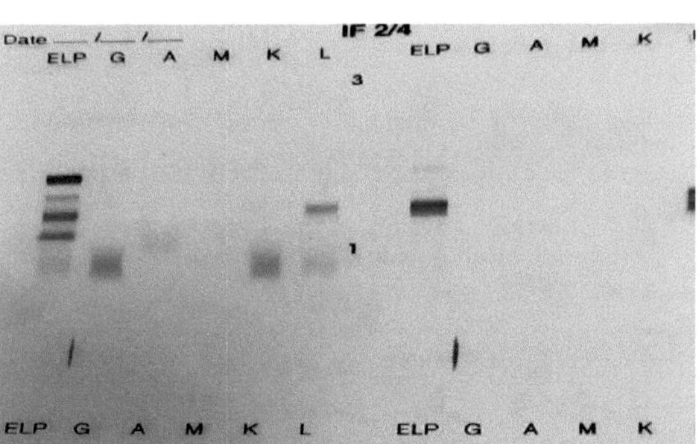 |
| • Profile oligoclonal : | 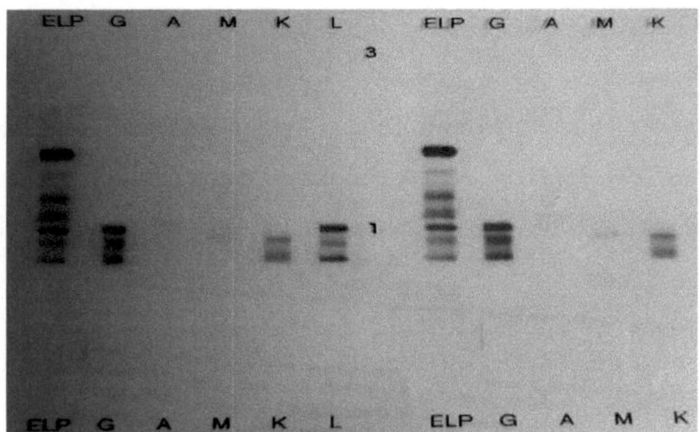 |

# CHAPTER 4
# Discussion

We studied cases of monoclonal gammapathy diagnosed in the clinical biochemistry department of the military hospital of Tunis over a period of 1 year.

## 1. Frequency of monoclonal gammopathies

In our cohort, the annual caseload of monoclonal gammopathies for the year 2019, from January to December, was 36 patients.

Numerous scientific researches on this pathology have shown an increase in the number of annual cases from 1995 onwards.

 Studies carried out in France: one in the internal medicine department of the Rennes University Hospital over a 16-year period (between 1990 and 2005) found an annual number of cases that did not exceed 20 cases per year during the period 1990 to 1995, whereas after 1995 cases of monoclonal gammapathies increased to 60 cases in 1996. (17)

Another study carried out at the Blois hospital center during the period 1980-2001 confirmed this increase starting in 1995, when cases rose from 50 cases in 1994 to 141 cases in 1995. (17)

These results can be explained by improved electrophoresis techniques.

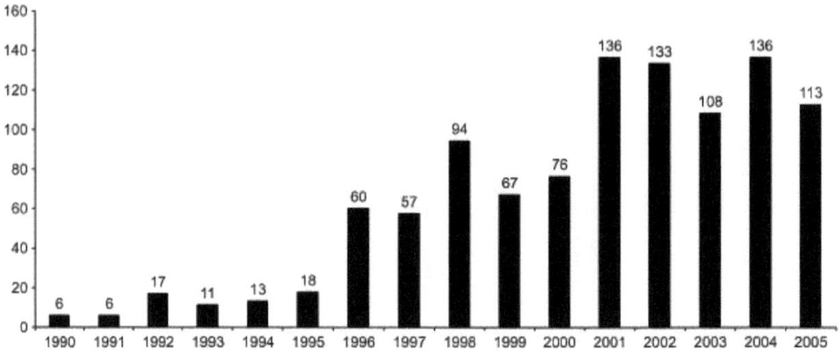

*FIGURE 12: REPRESENTATION OF THE ANNUAL NUMBER OF CASES OF DIAGNOSTIC MONOCLONAL GAMMAPATHIES IN THE INTERNAL MEDICINE DEPARTMENT OF CHU RENNES (17)*

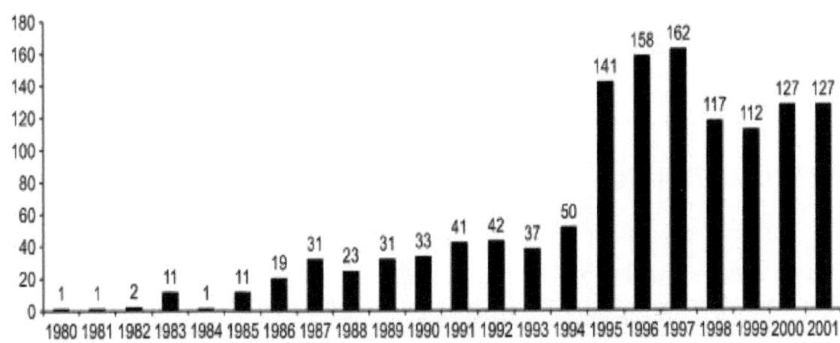


*FIGURE 13. **REPRESENTATION OF THE ANNUAL NUMBER OF CASES OF MONOCLONAL GAMMAPATHIES DIAGNOSED WITH CHU BLOIS (17)***

Our results show a predominance of monoclonal gammapathies in elderly patients where the mean age was 60 years with extremes ranging from 38 to 86 years.

These results are almost similar to those found by Decaux *et al.* (2007) in the internal medicine department of the Rennes University Hospital Centre during the period 1990 to 2005 with a median age of 71 years and in the Blois hospital center during the period 1980-2001 where the mean age was 79 years. No monoclonal gammopathies were diagnosed below the age of 30, which confirms our previous results. (17)

The first epidemiological studies concerning monoclonal gammapathies date back more than 30 years ago on geographically undefined populations and hospitalized patients but using less sensitive detection techniques than current electrophoresis. These studies have estimated a frequency of monoclonal gammapathies around 1% of the population.

In a systematic health examination carried out in Finistère, France, analysis of the sera of 3,0279 adults revealed 334 cases of monoclonal gammopathy (1.1% of patients)(19).

In 2006, **Kayle *et al*** conducted an epidemiological study to investigate the frequency of monoclonal gammapathies in the general population of Olmsted County, Minnesota in the United States. These researchers collected 2,1463 samples corresponding to

76% of the region's residents. The incidence of monoclonal gammapathies in this population of people over 50 years of age was estimated at 3.2%.

This frequency increased with age, ranging from 1.7% between 50 and 59 years of age to 6.6% over 80 years of age(20).

The pathology was significantly more frequent in men (350/9469 - 3.7%) than in women (344/11 994 - 2.9%), confirming our previous results where a male predominance was found with a sex ratio of 2:1.

## 2. Isotype of monoclonal gammapathies

### 2.1.      Heavy chains

*TABLE 9.ILLUSTRATION OF ISOTYPES*

| Isotype | Military Hospital of Tunis Number (percentage) | Internal Medicine Departme nt of Rennes Number (percentage) | CHU of Blois Number (percentage) | Comparison of results |
|---|---|---|---|---|
| IgG | 18 (50%) | 450 (42,8%) | 765 (59,7%) | concordant |
| IgM | 1 (2,78%) | 335 (31,9%) | 329 (25,7%) | Not concordant |
| IgA | 10 (28%) | 94 (8,9%) | 151 (11,8%) | Not concordant |
| IgD | 0 (0%) | 0 (0%) | 3 (0,2%) | concordant |
| Oligoclona l | 4 (11,11%) | 103 (9 ,8%) | 0 (0%) | concordant |
| Light chai n | 3 (8,33%) | 69 (6,6%) | 14 (2,7%) | concordant |

According to our results, IgG is in the majority, it is present in 50% of the cases of this pathology (18 patients). The distribution of isotypes was as follows: 50% IgG, 28% IgA, 2.78% IgM, 11.11% oligoclonal and 8.33% isolated monoclonal light chains. No IgD was found in our epidemiological investigation.

This is consistent with the data in the literature. However, IgA is more common in published studies. They accounted for 8.9% of cases in the Rennes internal medicine department and 11.8% of cases in the Blois hospital center. On the other hand, IgM is less represented in our work (2.78%) than that done by these two hospital centers. Indeed, the frequency of IgM monoclonal gammapathies was estimated at 31.9% in the Rennes internal medicine department and 27.5% in the Blois CHU(17).(17)

Monoclonal IgD or IgE gammopathies are rare. In our cohort, neither IgD nor IgE were found. No cases were reported in the study done at the Rennes internal medicine department, however 3 cases of IgD monoclonal gammapathies were reported at the CHU of Blois.

Thus, the frequency of light chains and oligoclonal profiles found in our cohort (8.33% for light chains and 11.11% for oligoclonal profiles) is practically the same estimated by the internal medicine service of Rennes (6.6% for light chains and 9.9% for oligoclonal profiles). Our results also agree with those of the CHU of Blois(17).

### 2.2.     Light chains

*TABLE 10.DISTRIBUTION OF LIGHT CHAINS*

| Light chain | Military Hospital of Tunis<br><br>Number (percentage) | Internal Medicine Department of Rennes<br><br>Number (percentage) | CHU of Blois<br><br>Number (percentage) | Comparison of results |
|---|---|---|---|---|
| Lambda | 20 (55,56%) | 702 (66,79%) | 788 (61,46%) | concordant |

| Kappa | 16 (44,44%) | 337 (33,21.%) | 492 (38,54%) | concordant |
|---|---|---|---|---|
| Kappa/Lambda ratio | 1,25 | 2.08 | 1.6 | concordant |

Regardless of the nature of the heavy chain, our retrospective study during the year 2019 showed a dominance of the Kappa light chain over the Lambda light chain. Kappa isotypes accounted for 55.56% (20 patients) of the cases in our cohort. Lambda isotypes accounted for 44.44% (16 patients) of the cases.

The kappa / lambda ratio was 1.25.

The comparison of our results with those found by the Department of Internal Medicine of Rennes and the CHU of Blois, confirms the hypothesis of dominance of Kappa light chains. Kappa light chains represented 66.79% (702 patients) of the cases in Rennes with a Kappa/Lambda ratio equal to 2.08. In Blois this Kappa/Lambda ratio is equal to 1.6.(17)

The difference in results found in our study with those concluded by the two French hospital centers for some parameters such as IgA and IgM frequency can be explained by the low number of patients and the short duration of our epidemiological study which lasted only 1 year and was insufficient to study some epidemiological parameters such as incidence. However, the internal medicine department of Rennes and the CHU of Blois carried out a study over a period of more than ten years and on a large number of patients: 1051 patients in Rennes and 1282 in Blois.

Urinary immunofixation allows us to highlight the presence of Bence Jones protein which was positive in 19 cases (52.78% of cases).

# CHAPTER 5
## Conclusion

Monoclonal gammopathy is a frequent pathology in daily clinical practice and this frequency is increasing.

In this work carried out at the biochemistry department of the military hospital of Tunis, we studied the epidemiological profile of this pathology. The results found show that monoclonal gammapathy is more frequent in elderly people with an average age around 60 years. A male dominance was raised.

This is consistent with the data in the literature.

On the other hand, typing by serum or urinary immunofixation found a dominance of IgG monoclonal gamapathies which are involved in half of the cases with a total absence of IgD and IgE which are rare.

For light chains, Kappa isotypes are the majority light chains in this pathology.

# Bibliographical references


1.  Vesterberg O. A short history of electrophoretic methods. Electrophoresis. déc 1993;14(12):1243-9.

2.  Vieira-Nunes AI. ION TRANSPORTATION UNDER THE EFFECT OF AN ELECTRICAL FIELD IN POROUS ENVIRONMENTS: APPLICATION TO THE SEPARATION OF RARE LAND BY FOCALIZING ELECTROPHORESIS [Internet] [phdthesis]. Institut National Polytechnique de Lorraine - INPL; 1999 [cited 11 Feb 2021]. Available at: https://tel.archives-ouvertes.fr/tel-00011215

3.  Godon Bernard, Loisel William. A Practical Guide to Analysis in the Cereal Industries / coordinators B. Godon and W. Loisel. Paris: Technique et documentation; 1984. xxiv+685. (Collection Sciences et techniques agro-alimentaires).

4.  Vigneron P. Principles of electrophoresis and its use in Forest Genetics. Bois For Trop [Internet]. 1984 [cited 11 Feb 2021]; Available on: https://agritrop.cirad.fr/444334/

5.  Memoire Online - Fractionation of serum proteins by electrophoresis on polyacrylamide gel - Hamza Belaouar [Internet]. Memoire Online. cited 11 Feb 2021]. Available at: https://www.memoireonline.com/07/09/2355/m_Fractionnement-des- serum-proteins-by-electrophoresis-on-polyacrylamide gel0.html

6.  Cotton F, Vertongen F, Gulbis B. Capillary electrophoresis and hemoglobinopathies. Immuno-Anal Biol Spec. Feb. 1, 2006;21(1):45-50.

7.  Schelcher F, Braun J-P, Guelfi J-F. Serum protein electrophoresis: principles of interpretation in dogs, cats and horses. Rev Médecine Vét. 1 Jan 1996;147:123 -30.

8.  Cellar CC, Lombard C, Dimet I, Kolopp Sarda M-N. Serum protein electrophoresis in medical biology: interferences and confounding factors. Rev Francoph Lab. 1 Feb 2018;2018(499):47-58.

9.  Emile C. Serum protein electrophoresis: general principles - method verification - interpretation. Option/Bio. Jan 1, 2013;24(483):20-2.

10. Le Garff-Tavernier M, Masmoudi S, Musset L. Use of Neurosoft® software as an aid in the interpretation of serum protein electrophoresis. Immuno-Anal Biol Spec. June 1, 2008;23(3):153-8.

11. Baudin B. Nephrotic syndrome. Rev Francoph Lab. 1 Sept 2013;2013(455):51-6.

12. Zandecki M, Geneviève F, Jego P, Grosbois B. Monoclonal gammapathies of undetermined significance. Rev Internal Medicine. 1 Dec 2000;21(12):1060-74.

13. Norbert I. Gammapathiesmonoclonalesde signification indéterminée.La Revue du Praticien56 :18-24. Rev Prat. 2006;(56):18-24.

14. Ghrairi N, Bouakkez H, Dahmouni A, Nahdi I, Mechmeche L, Boughnim L, et al. Difficulties during serum immunofixation. Immuno-Anal Biol Spec. 1 Apr 2009;24(2):100-3.

15. IMMUNOFIXATION.pdf [Internet]. cited 14 Feb 2021]. Available at: https://www.eurofins-biomnis.com/referentiel/liendoc/precis/IMMUNOFIXATION.pdf

16. Dasgupta A, Wahed A. Chapter 22 - Protein Electrophoresis and Immunofixation. In: Dasgupta A, Wahed A, éditeurs. Clinical Chemistry, Immunology and Laboratory Quality Control [Internet]. San Diego: Elsevier; 2014 [cité 11 févr 2021]. p. 391-406. Disponible sur: https://www.sciencedirect.com/science/article/pii/B978012407821500022X

17. Decaux O, Rodon P, Ruelland A, Estepa L, Leblay R, Grosbois B. Descriptive epidemiology of monoclonal gammapathies. Experience in a general hospital center and in an internal medicine department of a hospital and university center. Rev Internal Medicine. 1 Oct 2007;28(10):670-6.

18. Netgen. RMS 201 [Internet]. Swiss Medical Journal. cited 14 Feb 2021]. Available at: https://www.revmed.ch/RMS/2009/RMS-201

19. Saleun JP, Vicariot M, Deroff P, Morin JF. Monoclonal gammopathies in the adult population of Finistère, France. J Clin Pathol. janv 1982;35(1):63-8.

20. Kyle RA, Therneau TM, Rajkumar SV, Larson DR, Plevak MF, Offord JR, et al. Prevalence of monoclonal gammopathy of undetermined significance. N Engl J Med. 30 mars 2006;354(13):1362-9.

## Summary :

Monoclonal gammopathy is a common biological abnormality in the general population with an estimated incidence of 3% above the age of 50 years and increasing with age.

The aim of this book is to study the epidemiological and biochemical profiles of monoclonal gammapathies and to explain the principle of the two techniques of their biochemical exploration: serum protein electrophoresis on agarose gel and immunofixation.

**Key words:** Monoclonal gammapathy, epidemiological profile, serum protein electrophoresis, immunofixation.

## Abstract:

Monoclonal gammopathy is a common biological defect in the general population. The incidence is estimated at 3% above the age of 50 and the frequency continues to increase with age.

The aim of the work was to study the epidemiological profile of monoclonal gammopathy by dividing the cases found according to various criteria: sex, age and type of immunoglobulins. Carrying out this work requires the integration of two biochemical analysis techniques : Serum protein electrophoresis on agarose gel and immunofixation.

**Key words:** Monoclonal gammopathy, epidemiological profile, serum protein electrophoresis, immunofixation

Buy your books fast and straightforward online - at one of world's fastest growing online book stores! Environmentally sound due to Print-on-Demand technologies.

Buy your books online at
**www.morebooks.shop**

Kaufen Sie Ihre Bücher schnell und unkompliziert online – auf einer der am schnellsten wachsenden Buchhandelsplattformen weltweit! Dank Print-On-Demand umwelt- und ressourcenschonend produziert.

Bücher schneller online kaufen
**www.morebooks.shop**

KS OmniScriptum Publishing
Brivibas gatve 197
LV-1039 Riga, Latvia
Telefax: +371 686 204 55

info@omniscriptum.com
www.omniscriptum.com